REPORT

ON THE

OF

DODGE AND WASHINGTON COUNTIES

STATE OF WISCONSIN,

BY

JAMES G. PERCIVAL,

STATE GEOLOGIST.

Milwaukee:

STARRS' BOOK AND JOB PRINTING OFFICE, JUNEAU BLOCK.

1855

REPORT

ON THE

OF

DODGE AND WASHINGTON COUNTIES

STATE OF WISCONSIN,

BY

JAMES G. PERCIVAL,

STATE GEOLOGIST.

Milwaukee:

STARRS' BOOK AND JOB PRINTING OFFICE, JUNEAU BLOCK.

1855

REPORT

ON THE

IRON OF WASHINGTON AND DODGE COUNTIES.

The Iron Ore, at Iron Ridge, Dodge Co., and Hartford, Washington County, is a red peroxyd of iron, (of the same species as specular iron and red hematite,) chiefly of the variety called lenticular ore, (seed or shot ore.) It forms a bed interposed between two limestone formations; that below obviously corresponding in its physical character and fossils to the upper shell bed of the Blue Limestone of the Mineral district, and that above in its physical character to the Upper Magnesia of the same district. Fossils are very rare in this overlying rock, and those chiefly corals, near the junction of the rock with the ore. They are quite different from the fossils in a corresponding situation in the Upper Magnesia, so as to render the determination of the overlying rock not immediately obvious. The ore bed has the same relative position in regard to the underlying rock as the *brown* and *green* rocks of the mineral district, which are situated at the base of the Upper Magnesia, immediately above the Blue Limestone. Those rocks are highly stained with iron, particularly the brown rock, which has nearly the color of the present ore, and accompany the lower mineral openings of the Upper Magnesian, in which the lead ore is associated with large quanities of iron ore, (sulphuret of iron, and brown hematite; the latter derived from the decomposition of the former.) The iron ore, at the two localities above mentioned, is the same in its character and arrangement, and has the same geological relations. It forms, in its original position, a bed of thin slaty rock, composed of small flattened smooth grains, with some larger connections intermixed either of a mamaellary form, or resembling very smooth rolled pebbles, but these obviously concretionary. The bed is generally very uniform in its character, but with occasional thicker and more compact layers. It is usually overlaid by a layer, 3 to 4 inches thick, of a very hard dark form, blue seamed compact ore, breaking with a conchoidal fracture, with occasional glossy seams of specular iron. This layer in some instances adheres firmly to the lime stone above, and points of the same ore or stains of the red oxyd are found more or less disseminated through the adjoining rock. Iron pyrites is also found disseminated through the immediately overlying lime stone, and may be observed in the same block apparently passing from the unaltered pyrites at the centre to red oxyd at the surface of the block. At Hartford irregular layers or pockets of red and white jointed clay, blended in larger and smaller segregations or patches, are formed in the lime stone a few feet above the ore, very similar to the joint clay of the openings in the mineral districts, particulary those in its lower bed of the Upper Magnesian. This clay breaks by smooth seams into more or less regular fragments, sometimes very small, like those of the soap clay immediately investing the lead ore in some openings in the mineral districts. The rock adjoining these pockets of clay contains an unusual number of fossils (corals,) like that immediately overlying the ore. The ore is underlaid by a bed of red and blue clay, accompanied with fragments of a greenish concretionary lime stone, with few or no fossils resembling similar concretionary layers in the lower part of the Upper Magnesian, and this by the upper shell bed of the blue lime stone, usually much decomposed and broken, and accompanied with alternate layers of blue clay as in the mineral district.

At Iron Ridge, the ore bed underlies a line of bluff of the overlying lime stone, about 30 feet high, extending 1 1-6 mile nearly North and South (N. 8° E.) This bluff is interrupted for about ¼ mile at the Mayville ore bed, and terminates abruptly both at the North and South; the limestone bearing around nearly from West to East. The height of the ridge from the valley Wes , is about 60 feet; the upper half, lime stone; the lower half occupied by the ore bed, and the underlying clay and blue lime stone. The lime stone in this bluff is thick bedded, hard and compact, of a very light grey colour, and burns, though with difficulty into a good white lime. The ore bed, where exposed by excavations as it underlies the rock, is composed of a very uniform mass of thinly schistose iron stone (rock ore,) of a light red brown colour, but giving a bright red powder, and made up chiefly of very small flattened grains (seeds) of argillaceous red oxyd, with some larger smooth concretions disseminated. This is quite firm where formed in the pits sunk through the rock back of the bluffs, but softer in the excavations at the bluff, showing a tendency to disintegration. Where the ore meets the thick bedded rock above, a band, 2 to 4 inches thick, intervenes, composed of a very hard compact dark brown ore, giving a bright red streak and powder, and with occassional seams of specular iron. But in some instances a thin slaty marl is interposed between the lime stone and the ore, and in such cases the hard band appears to be wanting. At one of the excavations in the face of the bluff, a thin band of a brown ferruginous lime stone was found interposed in the ore, about six feet from its upper surface, thinning out towards the West, and becoming thicker as it passed under the lime stone.

On the slope of the ridge below the lime stone, and at the Mayville bed in the cove between the two sections of the lime stone bluff, and the South end of the ridge beyond the Southern termination of the lime stone bluff, the ore occurs loose and incoherent, but composed of the same small flattened grains, as the rock ore, with a few larger concretions intermixed. The ore here is arranged in layers, but less regularly than in the original bed, with more or less clay intermixed, both in horizontal and vertical seams, and with interposed irregular beds and pockets of bluish joint clay and a yellow brown loamy drift with boulders of lime stone; the whole preventing the appearance of a drift accumulation. The lime stone and the underlying ore may be supposed to have originally extended farther West, and to have been removed by the action of water, and the rock ore to have been disintegrated and then accumulated by eddies in the cove at the Mayville bed, and at the South end of the ridge, and to have been deposited by more gradual action, along the general slope of the ridge. This view of the subject will explain the great variety in the thickness of the loose seed ore at various points along the ridge; at the Mayville bed, including the interposed beds of clay and drift, about 30 feet, and at the summit of the South point of the ridge, about 20 feet, while on the slope of the ridge it gradually thins off from 8 to 10 feet above, towards the bottom.

Pits have been sunk through the rock at different distances from the bluff, and show the same arrangement of the bed as in the excavations in the rock in the face of the bluff, only less altered and disturbed. The thickness of the bed under the rock, in the South part of the ridge, appears to be about 15 feet. In one pit it was found to be 12 feet, but here the ore or the clay beneath, appeared to have been removed by the action of a spring, causing a sinking of the rock at that point. At the North end of the ridge, in the bluff North of the Mayville bed, the thickness of the bed under the rock, averages 10 feet (in one pit 12 feet.) A higher bluff of lime stone extends along the West face of the ridge, from two to three miles North of Iron Ridge village, where two pits have been sunk below the rock of the bluff, showing a mere seam of the iron ore under the overlying lime stone, below which, in one of the pits, the same clay and blue limestone were found as under the ore bed at Iron Ridge. The lime stone bluff is there more than twice as high as that at Iron Ridge, and shows beneath the same thick bedded rock as at the latter, overlaid by a bed of nearly equal thickness of a thinner bedded lime stone breaking into small jointed fragments, and this by another bed of thick bedded rock at the summit. The rock throughout is there as little fossiliferous as at Iron Ridge. The arrangement in these distinct beds corresponds with that common to all the lime stone formations of the mineral district.

At Hartford the ore bed crops out under earth, on the West slope of the ridge at the village, South of the Rubicon, and in the South bank of that stream, and has been traced in wells and pits through the ridge to its East base adjoining the Rubicon. It is overlaid on the West, first by earth, then by a very thin bed of limestone in places, but disjointed; and farther East by a firm bed of the same rock, 6 to 8 feet thick, under about 20 feet of earth, bu again appears on the East slope as on the Western. It has been sunk through only in one pit, towards the West, under the disjointed rock, and is there 7 feet thick, but thins out on both slopes, particularly on the Eastern. The ore appears, here harder than that at Iron Ridge, The loose seed ore is found only in comparatively small quantities on the West slope of the Ridge where the ore is separated from the limestone above by the same hard band as at Iron Ridge, presenting similar seams of specular iron. Pits have been sunk in the Ridge next East, where there is the same appearance of thinning out to a seam as in the high bluff North of Iron Ridge. These two instances indicate a thinning out both to the North and East, and the same if opportunity offered might probably be shown to the South and West. No discoveries of this ore have yet been made, except those at Iron Ridge and Hartford. The thin layers of red earth containing a few grains of the ore disseminated found on the surface of the drift at a few points South-east of Iron Ridge are undoubtedly due to the action of water and are derived from that bed. It would seem probable that the ore forms extensive lenticular deposits or basins, thinning out around their edges, and occurring at different points along the junction of the two limestone formations between which it is included. The thickness of the ore would naturally differ in different deposits, according to their extent. Thus the deposit at Iron Ridge which has been traced at least 1¼ miles North and South, and nearly the same distance East and West, shows a thickness of 15 to 16 feet under the limestone rock, while that at Hartford, which has not been traced to half that extent, shows a thickness of only seven feet. The available extent of the Hartford bed, on the North, is limited by the Rubicon, which runs from East to West, along the line of an apparent fault; the ore bed rising several feet above the level of the stream on the south, and the limestone presenting the same characters as the overlying limestone on the South, sinking below the level of the stream on the North.

From the statements above it will be seen that the ore available in quality occurs under two forms; the rock ore, its original form as it lies in place under the limestone, and the loose seed ore which has been apparently formed from the disintegration of the rock ore, and more or less modified in its deposition by drift action. The composition of the two cannot essentially differ, except that the latter is more or less intermixed with clay, and may have been modified by the action of decomposed organic matter, as in bog ore. A careful analysis of the two would determine the latter question. The loose ore is more easily excavated, but the rock ore would be reduced with greater facility, from its coherence. The hard band immediately adjoining the limestone above, although probably of superior quality, is in too small quantity to be of any importance, and is interesting only in pointing out the character of the ore.

The bed at Iron Ridge occupies an extent of nearly one mile from North to South, (the Mayville bed not included,) along the side of a ridge where the bottom of the bed is above the base, so that the whole bed admits of easy drainage. The thickness of the bed under the rock varies from 10 to 15 feet; at the south point, (at the village,) where the furnaces are to be erected, about 15 feet. By placing the furnaces at the side of the bluff at the thickest part of the bed, the ore may be brought to them immediately, almost without labor. The large deposit of loose ore at the south end of the ridge, (in one point 20 feet thick,) is in the immediate vicinity, and the ore could be conveyed to the furnaces with nearly equal facility. Either of the two might thus be used, or the two might be mixed as would be found most advantageous. But the rock ore would most probably be used to most advantage, and with its known extent would be inexhaustible.

This great deposit of ore is fortunately in the midst of a very extensive tract of heavily timbered country, which with due economy might for a long period furnish an abundant supply of fuel, and that of the best kind for the furnace, as the superiority of charcoal iron is acknowledged. The whole face of the ridge presents a series of springs, issuing both above and below the ore, and affording an abundant supply of water for steam power. The bed at Hartford is in the same heavily timbered country, and on the immediate bank of the Rubicon, a large mill stream. The Milwaukee and Horicon Railroad passes through Hartford Village, and only 1¼ miles from the village at Iron Ridge, and will be connected with the latter by a branch leading directly to the works. The surrounding country is of a superior character for agriculture, and will furnish abundantly all necessary supplies. It will thus be seen that these localities particularly that at Iron Ridge, furnish the most desirable advantages for the manufacture of iron, and by the connected Railroads for the transportation of the iron and the ore, from its abundance besides supplying the most extensive establishments on the spot, might be conveyed to other points, more or less remote, where it might be required, and that with great facility from the conveniencies of transportation, and thus every advantage might be taken of this vast deposit.

JAMES G. PERCIVAL.

From the Milwaukee Daily News, May 22nd, 1855.

Iron in Dodge and Washington Counties---James G. Percival's Report.

We alluded a few day since to a report of the State Geologist, relating to the large and valuable deposit of Iron Ore at Iron Ridge, with the remark that we should give it a more extended notice soon. We have read that paper with great interest, and are fully convinced that there is, slumbering at our threshold, one of the most important elements of wealth which exists within the boundaries of the State.

This body of ore extends over an area equal to about a mile and a quarter square, and at an average depth of about fifteen feet, which it is obvious will be inexhaustible for many centuries. It is said to smelt with a smaller amount of fuel by 25 per cent. than any ore known in the United States, and works with equal ease in the puddling furnace, producing the first quality of bar iron. This ore bed is situated in one of the heaviest timbered portions of the State, the predominant timber being oak and maple, or sugar tree, which are found to produce coal of the first quality, and in such an abundance as will admit of no scarcity for thirty years to come, and from which a supply of coal can be drawn for a much longer time at a reasonable cost. Before the surrounding forests will be exhausted, some substitute will doubtless be found to supply the place of charcoal, and thus the supply of mineral be made available from these mines for centuries.

There are but few deposites of iron to an equal extent in the United States, and none of those which have been discovered can be made so productive at so small cost as this at Iron Ridge. The iron from the Iron Mountains of Missouri and Lake Superior, is quarried and smelted with a much greater expenditure of labor and capital, and the iron resulting, is of no better quality when obtained. The Iron Ridge mines are in the immediate vicinity of the most beautiful and productive agricultural region of the State, whence supplies of provisions, labor, and all that is needful can be drawn at the cheapest rates, and in the most abundant quantities. The ore from the mines of Lake Superior and Missouri delivered at the furnace will cost some three dollars per ton, or about six to seven dollars for the amount of ore required to make a ton of pig iron; while the ore at Iron Ridge is obtained with such facility that it can be delivered from the mines to the furnaces erected at the foot of the ledge for TEN CENTS PER TON, or a sufficient amount of ore to make a ton of pig iron, can be delivered at the furnace for TWENTY-FIVE CENTS!!! Such a fact is unparalleled in the history of mining, but is believed to be a sober reality at Iron Ridge.

Thus the vast expenditures required for working the Lake Superior mines, remote from settlement and in an inhospitable region, where labor and provisions must always be expensive, are in a great measure avoided or greatly reduced, which in the large operations of a heavy iron establishment, would amount to the saving of thousands, and perhaps hundreds of thousands of dollars annually. We look upon this as one of the most valuable treasures of our State, greatly superior in its effects upon our domestic industry and wealth, to the richest gold-mines, or the most extensive veins of lead and copper. These are all valuable in their way, but none of them add so much practically to the permanent wealth and improvement of the country in which they are situated as do iron works in favorable locations. Such works not only furnish employment for a large number of persons in the various departments, but the production of their labor is an article of primary necessity in all the arts and business of life, and the demand for it is unlimited, as the supply of the raw material is inexhaustable.—These mines are situated on the La Crosse and Milwaukee Railroad only forty-five miles from this city, which will afford a ready means of communication and a market east and west by the navigation of the Lakes, to all the cities and ports holding the key to the markets of Illinois, Indiana, Michigan, Ohio, Pennsylvania, New York, and Canada; and by the Mississippi, to the western and north-western regions beyond us, not forgetting the large market which the State of Wisconsin itself will furnish, more valuable perhaps than all the others put together.

We understand that our fellow citizen the HON. BYRON KILBOURN, late Mayor of this city, is the principal proprietor and owner of this great body of mineral, and that he offers great inducements for the investment of capital in the erection of works and entering into the manufacture of Iron, both in the form of cast and bar Iron. For this purpose we are informed a charter has been obtained from the legislature with ample powers and guaranties, and the opening and opportunity for the investment of capital are such as seldom offer. All the circumstances considered, embracing locality—cheapness of obtaining and smelting the ore—facilities in obtaining supplies and convenience of market—there is probably not in the United States another locality which holds out stronger inducements for the investment of capital on a large scale, than the Iron region referred to.

www.ingramcontent.com/pod-product-compliance
Lightning Source LLC
LaVergne TN
LVHW011126110826
845150LV00008B/2260

* 9 7 8 1 4 1 8 1 9 5 5 8 8 *